Fizzlebop

Sermon on the Mount

5 fizztastically fun experiments and devotions for families

ISBN 978-1-946692-41-2 (pbk)
For Worldwide Distribution 1 2 3 4 5 6 7 8 9 10

For the FIZZTASTIC team at Tyndale Kids who brought Dr. Fizzlebop to families everywhere.

Making a Bee Home

The Beatitudes

(Matthew 5:3-12)

Introduction

You may not realize it, but you've been experiencing the stupendous benefits of bees all your life. From the trees and flowers growing outside your home to the fruits and vegetables you eat, bees have been involved in helping them to grow—and grow abundantly. But have you ever said thank you? I suppose I haven't either, at least not in a way they could understand. Hey, perhaps that's an idea for a future invention—B-Phone or B-Talk . . . The name might need some work as well as the science. In the meantime, I thought we could show our exuberant appreciation by building a home for bees. Interested?

WARNING! This experiment requires scissors. Safety first!

Fizzlebop Supply List

- Paper straws (or construction paper, pencil, and masking tape)

FIZZ TIP: Don't use plastic straws.

- Scissors
- Empty cardboard milk or juice carton
- Measuring tape
- Duct tape (variety of colors)
- 3 12-in (30 cm) zip ties
- Variety of colorful stickers

The Experiment

1. If you are using construction paper instead of paper straws, create the bee tubes by rolling the construction paper around the pencil. Use a piece of tape to keep the tubes from unrolling.

2. Cut off the top of the carton, creating an open square.

3. Measure the height of the carton. Then cut the construction paper tubes or paper straws to be slightly shorter than the carton's height.

4. Wrap the colorful duct tape around the outside of the carton to create a layer of protection from the weather. Add stickers for the finishing touch!

5. Fill the carton with the bee tubes, facing upward so the openings are accessible to the bees.

FIZZ TIP: Be careful not to bend any of the tubes, but make sure the carton is tightly packed. None of the tubes should fall out if you turn the carton upside down.

6. Find a tree branch where you can secure your bee home.

FIZZ TIP: Look for a spot that's secluded—not in an area where your family is usually active.

7. Wrap the zip ties around the bee home to secure it to the tree branch. These zip ties will keep it steady.

8. Over the next month, observe from a distance to see whether any bees have taken up residence.

"Dr. Fizzlebop, what's happening?"

Though our focus today was not an experiment but a service project, there is a bit of science involved in the packing of the carton. We didn't use tape or glue to hold the bee tubes in place—we only needed pressure and friction. The tighter we packed the tubes, the more pressure was exerted between and among the tubes, holding them in place. Friction between the tubes and the carton also helped to hold them in place.

FIZZ FACT: From the day honeybees are born, they zealously labor to pollinate as many flowers as they can. Did you know that every pound of honey the honeybees make represents over 2.6 million flowers pollinated? The honeybees in a hive will have traveled a distance equivalent to more than two times around the world to make that delicious treat. God designed honeybees to help our wondrous world have enough food to sustain life on planet Earth. In fact, honeybees do so much for our world that one in every three bites of food that you eat comes from a honeybee.

Devotional

I think we can all agree that the humble honeybee is a great model of a true servant. And bees are happiest when they are doing the work God designed them for. It's amazing how much impact a joyful servant's heart can have on the entire world.

Now, who else comes to mind when you think of a selfless and joyful servant like that? Jesus, of course! So it really is no surprise that Jesus preached about this very topic.

Have you ever heard of the Sermon on the Mount? In Matthew chapters 5 through 7, Jesus preached to a multitude (which means a whole lot) of people. Matthew 5:3-12 in particular is known as the Beatitudes (which is a list of blessings spoken by Jesus). Sensing a little tie between the experiments and our devotional—BEE-atitudes? Sometimes I can't help myself. Can you guess what the Beatitudes are all about?

As you read the passage, you'll find that all eight Beatitudes describe a joy-filled servant. Jesus starts each one with the phrase "blessed are"—and in this context, *blessed* really means "joyful" or "happy." There's a promise, a challenge, and a reward wrapped up in each one. Let's take a quick look at all eight of them.

1. **Bee joyful** when you are *poor in spirit,* because you will find that *your riches are in God's Kingdom.*
2. **Bee joyful** when you *feel you have lost what is most precious to you,* because you will find *the deep love of the one who is truly most precious.*

3. **Bee joyful** with *however little you have,* because you will find *the Lord is the one who provides everything you truly need.*
4. **Bee joyful** when you are *hungry for righteousness,* because you will be *satisfied.*
5. **Bee joyful** when you *care for others,* because you will find that *you have a heavenly Father who cares for you.*
6. **Bee joyful** when *your heart is right with God,* because you will find that *God is at work in the world around you.*
7. **Bee joyful** by *helping others get along,* because you will find *the peace that comes from being part of the family of God.*
8. **Bee joyful** even when *others treat you badly* because you followed God's example, because your golden reward—your honey that lasts forever—*will be great in heaven.*

Being a joyful and selfless servant isn't always easy, but the good news is that when the Spirit lives in us, he is able to change our hearts.

And so today as we think about some of God's most selfless servants, **bee joyful** that he has designed the amazing world in which you live—from the bee to the bison to the tadpole to the tiger. They've all got a place in this amazing world, and so do you!

Collecting Your Family's Data

Describe the location you chose to hang your bee home. Why do you think it appeals to bees?

____

____

Record the first day you noticed bees moving into the bee home.

____

____

Are there any Beatitudes that require some extra effort for you?

____

____

What are some ways you can use the Beatitudes in your daily life?

____

____

Prayer Time

Take a minute today to thank Jesus, the Magnificent Son, for being the perfect servant. Ask him to change your heart into that of a joyful servant so you can bless the world and spread his joy just like the honeybee spreads pollen and helps plants grow. Be sure to thank God for the honeybee, too!

Light-Up Gelatin

Light of the World

(Matthew 5:14-16)

Introduction

Gelatin: squishy, jiggly, wiggly, and yummy! Today we're going to watch our gelatin wiggle and… glow! Yes you read that right, glow! We've got a neato experiment today with a lot of scientific principles involved: states of matter, light, chemical reaction, reflection and more! WOW! It's going to be a wondrous experiment. Are you ready?

WARNING! This experiment requires a stove and boiling water. Safety first! Have an adult assist you.

Fizzlebop Supply List

- Packets of flavored gelatin
- Tonic water
- Cold water (optional)
- Stove
- Small pot
- Measuring cups
- Heat safe Bowl
- Wooden spoon or spatula
- Glass jar
- Fluorescent black light bulb

The Experiment

1. With an adult's assistance, boil one cup of tonic water in a small pot on a stove.

2. Pour the gelatin into a bowl as the tonic water heats up.

3. Once the tonic water boils, use a pot holder and an adult's help to add the tonic water to the bowl with the gelatin.

4. Dissolve the gelatin in the hot water by stirring it with the spoon or spatula.

5. Add one cup (total) of either unheated tonic water, cold water, or a combination of both to the bowl.

FIZZ TIP: The gelatin will light up better but taste more bitter if you use tonic water instead of plain water. Try different amounts and compare your results!

6. Transfer your gelatin mixture to the glass jar and let it solidify in the refrigerator.

7. Once the gelatin is solidified, set up your black light in a dark room and place your jar of gelatin underneath it.

8. Observe the luminescent appearance of your glorious gelatin!

FIZZ TIP: Experiment again to see what results other gelatin colors combined with more or less tonic water produce!

"Dr. Fizzlebop, what's happening?"

Gelatin contains a type of protein called collagen. The process of dissolving the gelatin mix in hot water followed by cooling it in the refrigerator caused the collagen proteins to bind together. The water and other tasty ingredients in the gelatin mix became trapped within the proteins' bonds as they solidified.

There are multiple forms of light, some of which can't be seen by the human eye. Your black light emits a particularly high-energy light form known as "ultraviolet." A bitter-tasting chemical in the tonic water, quinine, soaked up the energy from your black light and then released it almost instantly. However, the exchange reduced the amount of energy in the light and changed its form into one you could see. That's why the gelatin lit up!

Devotional

Jesus wants us to be like the gelatin in today's experiment. No, he doesn't want us to be soft and jiggly or deliciously edible. But he does want us to light up.

In the next part of his Sermon on the Mount, Jesus told his disciples, "You are the light of the world" (Matthew 5:14, NLT). Now, we aren't made of quinine, so how do we become lights? And why is that important, anyway? Open up your Bible to Matthew 5:14-16 and read what Jesus had to say.

This world can be a dark place, so Jesus wants his followers to shine with the light of God's goodness by doing good deeds! Remember all of the BEE-atitudes of a joyful servant from our previous experiment and devotional? When you live out the Beatitudes, you beat back a little more of the darkness with God's light. That allows people lost in the darkness to catch a better glimpse of God and to praise him for his goodness! Light is most clearly seen in the darkness, after all!

Sometimes, though, you may not feel like shining. You might not want to be noticed as different from everyone else. Maybe you see an opportunity to help your unsaved neighbor, but your friend invites you to come play a new game and says that everyone is going to be there, and you don't want to be left out or look weird by not going. Do you shine God's light by doing the unusual thing and serving your neighbor, or do you cover up that good impulse so you can blend in?

There's no way around it: living out the Beatitudes and shining God's light can be challenging. When it feels too hard, remember Jesus' promised rewards: You will find that your riches are in God's Kingdom. You will find the deep love of the one who is truly most precious. You will find the Lord is the one who provides everything you truly need. You will be satisfied. You will find that you have a heavenly Father who cares for you. You will find that God is at work in the world around you. You will find the peace that comes from being part of the family of God. Finally, your golden reward— your honey that lasts forever—will be great in heaven. I'd say that's worth lighting up over!

Collecting Your Family's Data

How long did it take your gelatin to solidify?

___

___

How much tonic water did you use? What do you think would happen if you tried more or less tonic water?

___

___

What color was your gelatin underneath the black light?

___

___

What do you think it would look like if you used a different color of gelatin?

___

___

How have you let your good deeds light up the world around you?

____

____

What are some good deeds you could do this week?

____

____

Prayer Time

Glorious God, thank you for your goodness. Please help me brighten up this world with good deeds. I want people to know and praise you! Amen.

A Bag of Rain

Love Your Enemies

(Matthew 5:43-48)

Introduction

Water, ice, gas—these are all forms of water. Today we're going to experience the water cycle. We'll be eyewitnesses as water turns to gas (evaporation) and back to liquid (condensation). Those might seem like big words, but in this super-simple experiment, you'll see them wondrously happen right before your eyes! God made this cycle to keep Earth working perfectly as planned. Magnificent!

FIZZ FACT: Did you know nearly 71% of the Earth's surface is covered with water, and did you know on average an adult human body is about 60% water?

WARNING! This experiment can involve high heat. Get an adult's assistance. Safety first!

Fizzlebop Supply List

- A gallon-size plastic bag
- Permanent marker
- Water
- Blue food coloring
- Packing tape
- A sunny window or hair dryer

The Experiment

1. Use the permanent marker to create some waves at the bottom of your bag. Draw a sun and some clouds at the top.

2. Fill the bag with about 2 cups or 500 mL of water and several drops of blue food coloring.

3. Zip the bag completely shut.

FIZZ TIP: Place packing tape over the seal for extra security.

4. Attach the top of the bag to a sunny window using a piece of packing tape. Observe how the water in the bag changes over the course of the day!

FIZZ TIP: If your window isn't sunny enough, try warming the bag with a hair dryer on its lowest setting. Be careful to not melt the bag! Keep the hair dryer a safe distance from the bag and don't aim it in one place for too long.

"Dr. Fizzlebop, what's happening?"

The plastic bag seals the water in an enclosed environment so we can observe the fizztastic phenomenon that is the water cycle! The heat from the sun and the hair dryer causes some of the water to evaporate into its gaseous state, also called water vapor. "Humidity" refers to how much water vapor the air is holding. When the air can't hold any more water vapor, the water vapor condenses back into water droplets which fall out of the air (or in our experiment, drip down the sides of the bag) as precipitation to start the cycle all over again! Marvelous!

Devotional

Think of someone who loves you. How do you treat them in return? It's easy to love them back and do good things for them, isn't it?

Now think of someone who's harder to get along with. Do you still treat them in a loving way?

So far, we've talked about lighting up the world by being a joyful servant. But you may be wondering, *Do I have to serve [insert name here]? They don't treat me very nicely! Why should I be nice to them?*

Not everyone in Jesus' day loved God or treated him very nicely, either. Some ignored his wishes; some didn't believe he existed; some probably made fun of him. But did God turn his back on those wicked and unjust people? Open up your Bible to Jesus' Sermon on the Mount and read Matthew 5:43-48 to find out!

Our experiment today illustrated how important sunlight and rain are to our planet's water cycle. Many of Jesus' listeners would have depended on that cycle of sunshine and rain to help their plants grow and keep their livestock alive—in other words, they themselves needed sunlight and rain to survive. Jesus explains in the passage you just read that God sends sunlight and rain to everyone no matter who they are or what they've done. God doesn't withhold his blessings from some people out of spite. He loves everyone equally. He loves you. And he wants you to love others with the same love he gives to you!

No, showing love to everyone isn't normal, and it isn't easy. (By the way, if someone is seriously harming you, showing love doesn't necessarily mean staying quiet and continuing to be hurt; asking a trustworthy adult for help could be the most loving thing to do for that person who's hurting you.) Do you remember our previous experiment and devotional about being a shining light?

Actively loving your "enemies" (the people who get on your nerves) by joyfully serving them through doing good deeds shines way brighter than only loving your friends because it's so rare! It's one of the best ways to show the world what God is like.

Collecting Your Family's Data

How much time passed before you noticed the water vapor condensing on the side of the bag?

Did you see much precipitation in the bag?

How can you lovingly and joyfully serve someone you normally have trouble getting along with?

Prayer Time

Father God, thank you for loving and blessing me. Please help me to show love to the difficult people in my life this week. Amen.

Corroding Nails

Storing Treasures in Heaven

(Matthew 6:19-21)

Introduction

Have you ever found something old in an attic, a garage, or even buried in the dirt outside? Maybe it wasn't well protected from moisture, like humidity, or rain even. I once found what used to be a sewing machine down in a cellar. I have no idea how long it was there, but all but the metal had long deteriorated, and the metal that remained was rusted and corroded. Today we're going to see the effects of different levels of moisture on objects. What's wondrous about this experiment is how we use certain constants (which means items that remain the same) under different variables (which means items that are different) to see significantly (which means very) different results on the objects—in this experiment, nails and screws. Are you ready?

WARNING! This experiment requires a stove and boiling water. Safety first! Have an adult assist you.

Fizzlebop Supply List

- Five disposable bottles
- Permanent marker
- Uncooked white rice
- Water
- Salt
- Cooking oil
- 5 iron nails
- 5 galvanized screws
- Pot
- Stove
- Bowl

• Spoon
• Paper towels

The Experiment

1. With an adult's help, boil some water in a pot on the stove and let it cool completely.

2. Pour some unboiled water into a bowl. Add salt and stir with a spoon until the salt dissolves.

3. Use the permanent marker to label each of your bottles: air, rice, water + air, saltwater + air, and water.

4. Drop one iron nail and one galvanized screw into each bottle.

5. Pour some rice around the nail and screw in the bottle labeled, "rice."

6. Pour enough unboiled water into the bottle labeled "water + air" to cover the nail and screw.

7. Pour your saltwater into the bottle labeled, "saltwater + air," covering the nail and screw.

8. Use the boiled and cooled water to cover the nail and screw in the bottle labeled, "water." Pour a layer of oil on top of the water in the bottle.

   SAFETY WARNING: Make certain the boiled water has completely cooled off, especially if you are using a plastic bottle!

9. Let the bottles sit in a safe location for anywhere from a week to a month and observe them regularly to monitor any changes to the nails and screws! Take a photo or

write down a description of what you see in each bottle every time you check it.

10. When you're ready to conclude the experiment, spread out some paper towels. Pour the oil/water combination into something you can throw in the trash. Pour the plain water and the saltwater down the drain. Dump the nails and screws onto the paper towels.

FIZZ TIP: Either label your paper towels with a permanent marker (make sure the marker doesn't bleed through!) or position the screws and nails carefully so you can remember from which bottle each screw and nail came.

11. Observe and compare the differences between each of the nails and screws.

"Dr. Fizzlebop, what's happening?"

You likely noticed that the nails in the bottle of water + air and the bottle of saltwater + air changed the most. That's due to the process of corrosion!

Corrosion occurs when a metal, such as iron, comes in contact with water and with the oxygen in the air. The metal's reaction with those elements causes a type of compound called an "oxide" to form on the metal's surface. Contact with water makes the metal molecules change their structure and enables them to bind with oxygen and corrode! Adding an electrolyte, such as salt, to the water increases the rate at which the metal molecules' structure changes, which hastens the entire process of corrosion.

You didn't add water to the bottle of air, but you may still notice some rust on the nail from that bottle. Air can have varying amounts of water (this is called air's humidity) in addition to oxygen. The amount of rust you see on this nail will likely depend on the humidity of your air supply.

The rice helped absorb the water in the air in its bottle, leaving even less water to initiate the corrosion process.

Boiling the water helped to remove oxygen from the water. Adding a layer of oil above the boiled water with the nail helped seal out the oxygen in the air. This also hindered corrosion.

But what about the galvanized screws? You should see less corrosion on them compared to the iron nails. "Galvanized" means the screws were coated with a metal that corrodes very quickly. Once that top layer of metal corrodes, it shields the second metal underneath from further corrosion.

Devotional

What is your greatest treasure? Do you keep it someplace special where it will be safe? As you can see from this experiment, the place you keep something plays a huge role in how long it lasts and how well it's protected.

Jesus also taught about this topic in his Sermon on the Mount. Take a look at Matthew 6:19-21. Physical items that you can see, hear, taste, touch, and smell—like toys or clothes—can only be kept here on Earth. And because of sin in this world, they're going to change, like those nails and screws that corroded. Physical treasures won't survive this world forever.

But before you get too depressed (which means sad), re-read verse 20. So there is a safe place to store our treasures where they will last forever! But what kind of treasures can we store in heaven? We can't exactly pack up our stuff in a mailing box and ship it to an angelic post office!

Jump up to the beginning of Matthew 6 and read verses 1-4. Now, in one of our previous experiments and devotionals, we talked about letting our good deeds shine out into the world, but here Jesus says to not let anyone notice what we do. The key is to let our good deeds illuminate God, not ourselves. If we do good deeds for that reason, then God will reward us even if no one else does. And those are the rewards that are being stored in heaven for us! If you'd like to know what rewards God has for you, go back and read the promises given to the joyful servant who does good deeds by living out the Beatitudes. Pretty fizztastic, I'd say!

One final thought. Look back at verse 21 of Matthew 6. It's easy to treasure the stuff we have on Earth, isn't it? It's right here in front of us and demands our attention and focus—our heart. But when we accept Jesus as our Savior and Lord and begin to develop a relationship with Him, He changes us: not like corrosion changed the nails, but in the way that being galvanized changed the screws. The screws lived in the same environment as the nails. Some corrosion attached to the screws, sure, but they endured better than the nails. Jesus will help us through this world despite all the messy things in it as we look forward to receiving our heavenly rewards. Like this experiment took some time, it may be a while before you see God's rewards. But it will be worth it in the end!

Collecting Your Family's Data

When did you first see the corrosion appear?

____

____

What corroded first: A nail or a screw? Which bottle was it in?

____

____

How long did you leave the nails and screws in their bottles? What do you think would have happened if you left them in there for a longer or shorter amount of time?

____

____

What do you treasure most?

____

____

How can you store up treasures in heaven?

____

____

Prayer Time

Ask Jesus to "galvanize" you and protect you from being corroded as you live in this world. Ask him to help you treasure God's heavenly rewards more highly than the stuff of Earth. Thank him for what awaits!

Quicksand

Building Your House on the Rock

(Matthew 7:24-27)

Introduction

Imagine with me for a minute that we are walking through a thick jungle in our boots and khakis. We're on the path to discovering a fascinating new slug, believed to only live in this part of the wilderness. Suddenly you hold up an arm and stop me. "What's wrong?" I ask, and you point to a mysteriously open spot ahead of us where nothing grows. "It's quicksand," you say. Thankfully you recognized the hazard, so we skirt (which means go around) the quicksand and continue on to discover the unique slug, which we name by combining the first three letters of your first name with the last three letters of my last name. And in honor of our imaginary adventure, today at Fizzlebop

Labs we're going to look into the unique properties of quicksand by making some of our own.

Fizzlebop Supply List

- 1 cup (120 g) cornstarch
- ½ cup (120 mL) water
- Glass mixing bowl
- Spoon
- Plate

The Experiment

1. Mix the cornstarch and water in the glass bowl.

2. Observe and record what happens when you stir it quickly versus when you stir it slowly.

3. Observe and record what happens when you poke it with a spoon or press on it hard with your hands.

4. Observe and record what happens when you scoop it up with a spoon and drip it onto a plate.

FIZZ TIP: You can store this mixture if you want to. But always stir the quicksand right before you use it.

"Dr. Fizzlebop, what's happening?"

Quicksand appears to be liquid, but when you push on it, it becomes firm under pressure. When the cornstarch grains are stirred quickly, they have difficulty sliding over each other due to the lack of water between them. When you stir them slowly, more water can get between the grains, and they slide over each other more easily. The same happens with poking. If you poke quickly, the substance becomes very hard. If you poke slowly, it remains runny.

Devotional

As we wrap up our series of experiments and devotionals based on Jesus' Sermon on the Mount, I want you to imagine something with me. Imagine you're a grown-up, and you get to design a house and have it built exactly how you want it. What would you include in your house? Where would you build it? In your hometown? By the beach? In the mountains? By a lake? Would it be a cottage, a tree house, a castle?

No matter where you build, one thing is extremely important: the foundation. What do you think would happen if you built a house right on top of a sandy beach? Open your Bible to Matthew 7:24-27 to see if you guessed it.

So how do we make sure our house is built on solid rock? If you were building at the beach, you'd need to drill down into the bedrock beneath the sand and anchor the house to that deep, firm foundation.

But what about our lives? Every action we take, every choice we make, is like a brick in the building of our lives. So how do we make sure we're building on rock instead of sand? Hmmm . . . if sand represents things that change and shift and collapse in life, what are some examples of those?

Perhaps you've gone through a big change, such as who you hang out with or where you live. If you've ever moved or gone to a new school, you know what that's like. Perhaps a grandparent or someone close to you has died. Perhaps one of your friendships has ended. Whether your loss has been great or small, you know how hard it was. Sometimes our lives change dramatically, and we may feel as if our whole world is being shaken.

But would you like to know a wonderful secret? It's found in one of my favoritest verses, Isaiah 54:10. "'For the mountains may move and the hills disappear, but even then my faithful love for you will remain. My covenant of blessing will never be broken,' says the LORD, who has mercy on you."

Did you catch that? Even if the entire landscape changes, God's faithful love for you will remain! His marvelous, never-ending, steadfast love for you is as solid as a rock and is a firm foundation on which you can build your entire life. His love is also a safe place to put your hope! Of all the things in this world, nothing is as steadfast or as dependable as God's great love for you. Count on it!

Collecting Your Family's Data

Describe the quicksand when you stirred it quickly versus when you stirred it slowly.

____

____

Describe what happened when you poked it with a spoon or pressed on it hard with your hands.

Describe what happened when you scooped it up with a spoon and dripped it onto a plate.

What's the biggest change you've gone through in your life? How did it affect you?

When it feels like your world is being shaken, how can you hold on to God's great love?

____

____

Prayer Time

Amazing Creator, thank you for your steadfast, unshakable love. Help each of us to put our trust in you and only you—and to build our lives on your firm foundation. Amen.

Want more great faith-filled devotions and fizztastic experiments? Visit fizzlebop.com

Watch Dr. Fizzlebop and his lab assistants conduct fizztastic experiments on YouTube.

Discover 52 fizztastic experiments and devotions for kids!

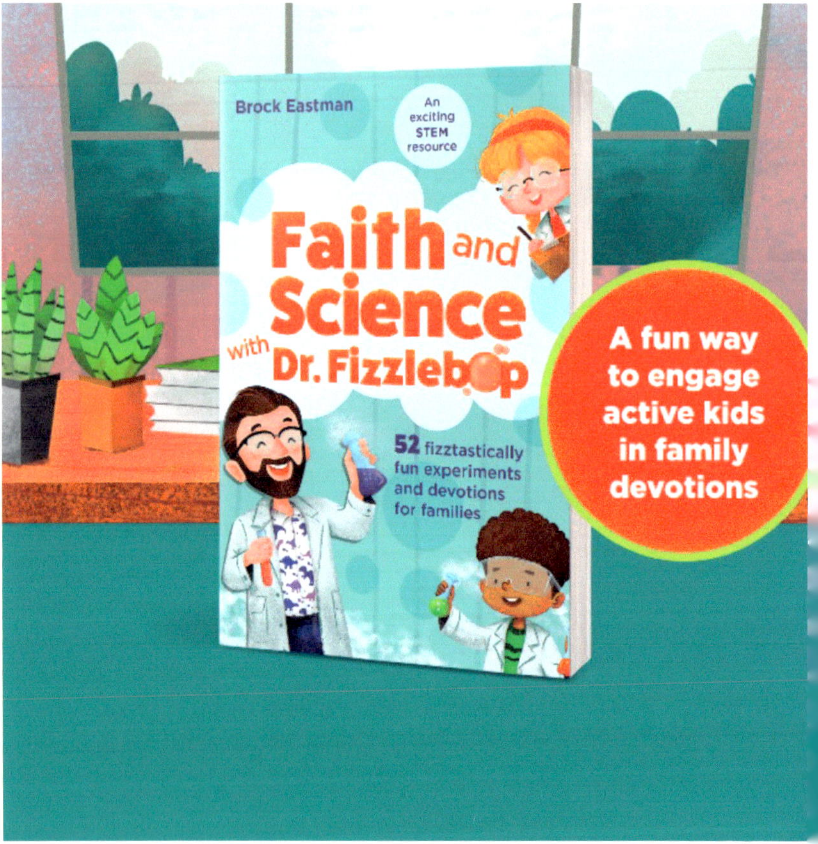

Meet the author!

Brock Eastman has a degree in Marketing and works for David C. Cook leading digital product and content development. Previously he was a producer and podcast host for *Adventures in Odyssey*. He writes feature stories and the Dr. Fizzlebop recurring column for *Clubhouse* and *Clubhouse Jr.* magazines. He is also the host of Tyndale Kids' Summer Camp. He loves cooking with his wife and reading stories to his four kids each night.

More books by Brock Eastman

Taken (P&R Publishing & Focus on the Family 2011)

Showdown with the Shepherd (Tyndale & Focus on the Family 2011)

Risk (P&R Publishing & Focus on the Family 2012)

Coming Storm (Crimson Pulse Media 2012)

Unleash (P&R Publishing & Focus on the Family 2013)

Tangle (P&R Publishing & Focus on the Family 2015)

Rainbow Hippo (Crimson Pulse Media 2015)

Count Hippo, Count (Crimson Pulse Media 2015)

Happy Hippo (Crimson Pulse Media 2015)

Alpha Hippo (Crimson Pulse Media 2016)

Wasted Wood (Crimson Pulse Media 2017)

Waste Deep (Crimson Pulse Media 2017)

Truthful Test (Crimson Pulse Media 2017)

Davey and the Mech (Crimson Pulse Media 2017)

Hope (P&R Publishing & Focus on the Family 2019)

Daddy's Favorite Sound (Harvest House Kids 2019)

Mommy's Favorite Smell (Harvest House Kids 2019)

HowlSage (Crimson Pulse Media 2020)

Bedtime on Noah's Ark (Harvest House Kids 2020)

Afraid of the Light (Crimson Pulse Media 2021)

Faith and Science with Dr. Fizzlebop (Tyndale Kids 2021)

Welcome to Hippopolis (Crimson Pulse Media 2022)

Obby the Obot (Crimson Pulse Media 2022)

Fizzlebop: Sermon on the Mount (Crimson Pulse Media 2022)

Fizzlebop: Explore the Senses (Crimson Pulse Media 2022)

Edge of the Galaxy (Tyndale & Focus on the Family 2023)

Made in the USA
Las Vegas, NV
09 November 2026

57757231R00036